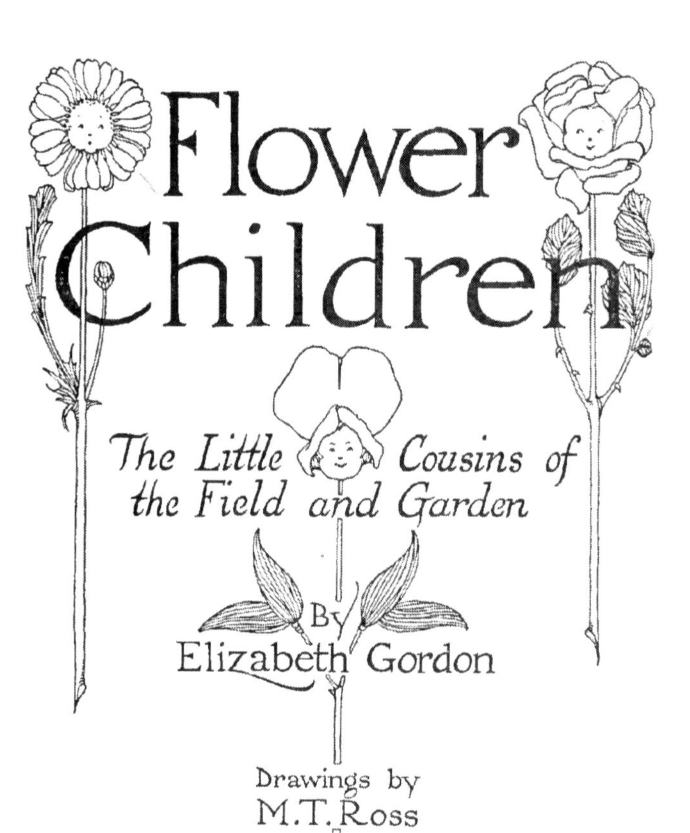

Flower Children

The Little Cousins of the Field and Garden

By
Elizabeth Gordon

Drawings by
M. T. Ross

Published by
P. F. Volland & Company
Chicago

Fifth Edition

To every Child-Flower that Blooms
Within the Glorious Garden
That we Call Home
This Little Book
is
Lovingly Dedicated.

FOREWORD

A flower, a child, and a mother's heart—
These three are never so far apart
A child, a flower, and a mother's love—
This world's best gifts from the world above

ALL children are flowers in the garden of God's love. A flower is the mystical counterpart of a child. To the understanding heart a child is a flower and a flower is a child. God made flowers on the day that He made the world beautiful. Then He gave the world children to play amid the flowers. God has implanted in the breasts of children a natural love for flowers—and no one who keeps that love in his heart has entirely forsaken the land of childhood.

In preparing this book the author and the artist have attempted to show the kinship of children and flowers, and it is their hope that the little ones into whose hands this volume comes will find herein the proof that their knowledge of what flowers really are is true and that their love for the friendly blossoms is returned many-fold.

To you, then, little child-flowers, this book is lovingly offered as an expression of thankfulness to children for the joy and sweetness with which they have filled my life.

—ELIZABETH GORDON

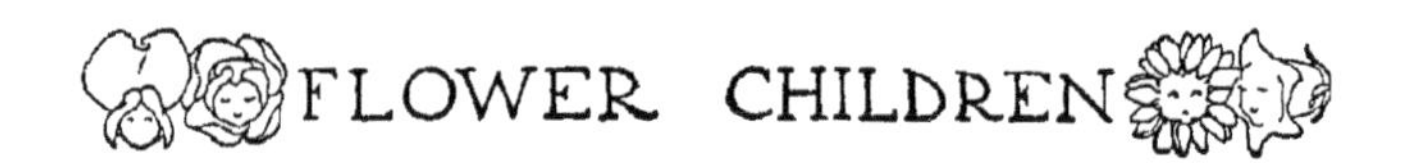

SAID CROCUS: "My! this wind is cold!
Most wish I had not been so bold,
Here the fields are still all brown;
Glad I wore my eider-down."

TRAILING ARBUTUS, you know,
Loves to grow beneath the snow.
Other folks would find it chilly,
She says that 's absurdly silly.

EAGER little Daffodil
Came too soon and got a chill;
Jack Frost pinched her ear and said,
"Silly child, go back to bed."

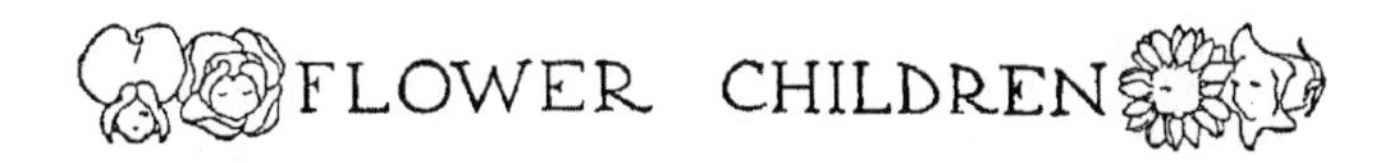

HYACINTH, the pretty thing,
Comes to us in early spring;
Says she always loves to hear
Easter bells a-ringing clear.

PUSSY WILLOW said, "Meow!
Wish some one would tell me how
Other kittens get around
And roll and frolic on the ground."

PRIMROSE is the dearest thing —
She loves to play out doors in spring;
But if a little child is ill,
She 's happy on the window sill.

GRANDFATHER Dandelion had such
pretty hair,
Along came a gust of wind and left his
head quite bare;
Young Dandelion generously offered him
some gold,
To buy a cap to keep his dear old head
from being cold.

WIND-FLOWER on an April day,
Came along and said she 'd stay;
Wore her furs snug as you please,
Said she liked the nice, cool breeze

ANEMONES and Bluets grew,
All the woodland pathway through;
Came along one day together,
Did n't mind the April weather.

LILAC wears a purple plume,
Scented with a sweet perfume;
Very high-born lady she,
Quite proud of her family tree.

TRILLIUM said "Why, deary me,
I 'm just as freckled as can be,"
Her cousin Tiger-Lily said,
"Well, look at me, I 'm almost red."

FLOWER CHILDREN

PANSIES like the shaded places;
With their little friendly faces,
Always seem to smile and say:
"How are all the folks to-day?"

JOHNNIE-JUMP-UP made a bet,
That he could pass for Violet.
What spoiled the little rascal's game?
The scent he used was not the same.

LADY'S-SLIPPER in the wood,
Said she really wished she could
Have a pretty dress and go
With sister to the flower show.

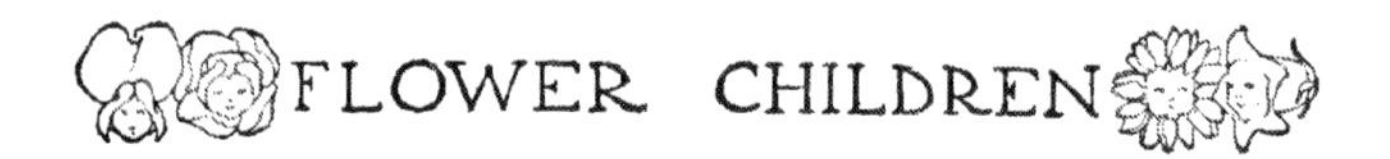

MODEST little Violet
Was her loving Mother's pet;
Did n't care to go and play,
Rather stay at home all day.

COWSLIP dearly loves to romp
Around the bottom of the swamp;
She comes along in early spring,
Before the grass, or anything

FRAGRANT little Mignonette,
In a shower got quite wet;
Laughed and said she did n't care—
It looked like jewels in her hair.

APPLE BLOSSOM is a fairy,
Swinging in a tree so airy;
By and by the little sprite
Sprinkles the ground with pink and white.

LITTLE golden-hearted Daisy
Told the sun that she felt lazy;
Said the earth was quite too wet,
She thought she would n't open yet.

LAUGHING, lucky Four-Leafed Clover
Is a most atrocious rover;
Does n't stay long in one place,
Goes and never leaves a trace.

THIRSTY little Buttercup
Caught the dew and drank it up,
Said cool water was so good,
She did n't seem to care for food.

SWEET little maid Forget-Me-Not,
She 's such a darling little tot;
A blue-eyed child with modest ways,
She 's never spoiled a bit by praise.

COMMON little Garden Pink,
Went away to school — just think!
When she came home for vacation,
Made them call her Rose Carnation.

BOUNCING BETTY stood all day
In the hedge row by the way;
By-and-by she crept outside,
And got so scared she nearly cried.

MORNING-GLORY thought she 'd look
Through the window at the cook;
Did n't know 't was impolite
To give a body such a fright.

HONEYSUCKLE, pretty vine,
Loved about the porch to twine.
Thought 't was just too sweet for words
To visit with the humming-birds.

WILD ROSE runs round everywhere,
Likes to breathe the nice fresh air;
Even her high-bred connection
Cannot match her pink complexion.

COLUMBINE 's a happy sprite,
Dances with fairies every night;
She feeds them honey when they go,
That 's why the fairies love her so.

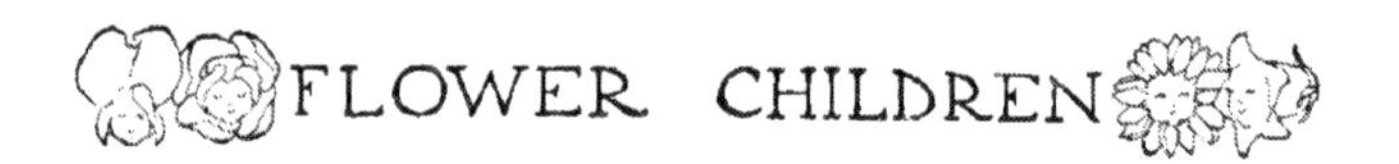

CUNNING LITTLE Blue-eyed Grass,
Smiles up at you as you pass;
Looks as if a bit of sky,
Had fallen down from 'way up high.

NASTURTIUM grew so big and tall,
He climbed up on the garden wall;
His little sister could n't go —
Dear child, she never seemed to grow.

PANSY SAID she wished she knew
What made Lark-spur look so blue;
Larkspur smiled and said 't was only
'Cause she felt a little lonely.

OH, HAVE you seen the sweet Briar-Rose?
She wears the very dearest clothes,
A hat the sweetest ever seen,
And dainty frock all shades of green.

BLUEBELL softly, gently sways
Through the long hot summer days;
Lives where nothing else can grow,—
That 's why we all love her so.

GERANIUM wears a scarlet gown,
With trimmings shading into brown;
Her cousin is a dainty sprite,
She dresses modestly in white.

SWEET ALYSSUM plays around
On any little piece of ground;
Takes up hardly any room,
And sheds a very sweet perfume.

SIMPLE LOOKING Blue-eyed Flax
Helped the farmer pay his tax;
Was busy all the season through;
Said it was n't hard to do.

BLEEDING-HEART, against the wall,
Told her woes to one and all.
Live-Forever said, "Forget it;
Life treats you the way you let it."

SWEET PEA said she thought they might
Give her a dress that was n't white;
So Mother Nature chose for her
All the colors that there were.

SNAP-DRAGON is so very bold,
He plays his tricks on young and old;
Hides behind the old stone wall,
And shoots his pop-gun at us all.

RAGGED ROBIN on a lark
Stole inside of Central Park;
There they treated her so well,
She soon looked like a city Belle.

YARROW PINK and Yarrow White,
Stole in on the lawn one night;
Gardener said they had no sense,
But they did n't take offense.

IRIS in a country garden,
Politely said, "I beg your pardon,
But I'm from sunny France you see,
And my real name is Fleur-de-Lis."

PEONY 'S a charming lady,
She does n't like a spot too shady;
Likes to live out in the light,
Dressed in red or pink or white.

ONCE THEY LOST sweet Babe Verbena,
Mother said, "Oh, have you seen her?"
But pretty soon the dear was found
Creeping on the nice soft ground.

DAINTY LITTLE Maidenhair
Lost her way and did n't care;
Played all day, the naughty child,
With common ferns, who run quite wild.

YOUNG Sweet William, sad to tell,
Rang the Canterbury's Bell,
"Just for that," his father said,
"William, come out in the shed!"

NAUGHTY little Four-O'Clock
Gave her mother quite a shock;
Stayed awake till nearly six,
Oh, she 's always up to tricks.

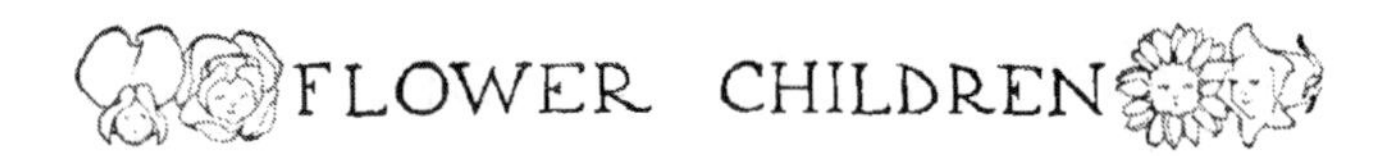

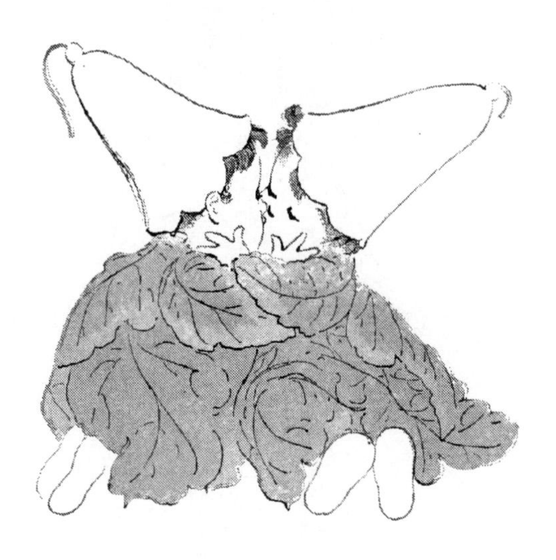

IF YOU 'RE very, *very* good
When you 're walking in the wood,
Twin-Flower babies you may see,
Sheltered by some old pine tree.

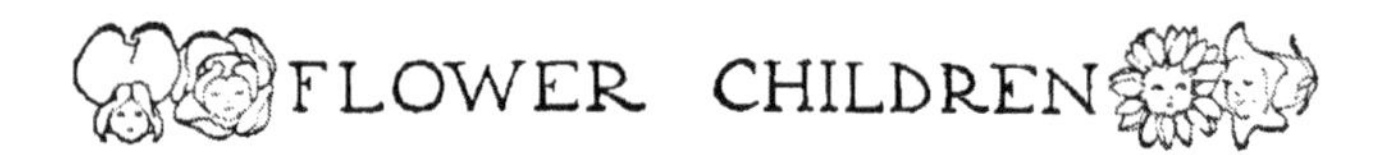

ALL THE SUMMER, Milkweed played,
Like a dear, good little maid;
But on a bright October day,
She found some wings and flew away.

CANDYTUFT and Marigold
Live outdoors until it's cold;
Sturdy maids with glowing faces
Blooming in the bleakest places.

JACK ROSE said, ambitiously,
He would grow to be a tree;
But his Dad said, "Better far
Be contented as you are."

WATER-LILY is very fond
Of floating in a sunny pond.
Tantalizing little creature,
Likes to grow where one can't reach her.

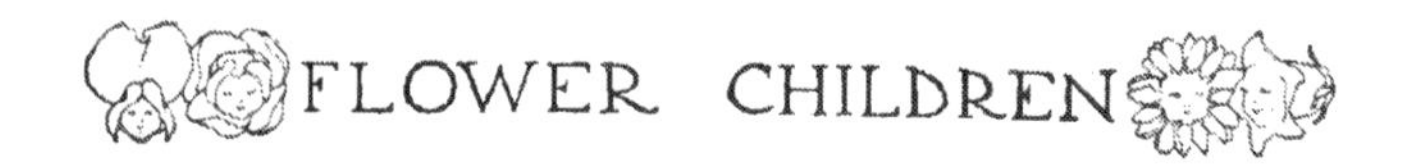

YOUNG COCKS-COMB was so very vain,
Hated to stay out in the rain;
Said he would n't so much care,
If he had other clothes to wear.

ON THE BORDER of the wood
All alone the Ghost-Flower stood,
Like a moonbeam dressed in white,—
Such a very pretty sight.

GOLDEN-ROD, the lucky chappy,
Grew up strong and tall and happy.
Slept out doors, if you 'll remember,
All those cold nights in September.

MADAME DAHLIA, like her name,
Is a very stately dame;
Her family is so polite,
It is a joy to meet them, quite.

MISS California Poppy said
She liked the sunshine on her head,
Though her friends might think her foolish,
Thought this country rather coolish.

CORN FLOWER, Bachelor Button's sister—
Gay young dog, he never missed her—
Went to live with Mrs. Corn,
So she would not be forlorn.

CAT-TAIL, growing in the marsh,
Thought his Mother very harsh,
Because she wouldn't let him play
With Blue Flag-lilies all the day.

PRINCE'S FEATHER, straight and tall,
Grew against the garden wall;
Did n't care to play, said he
Came of a royal family.

CRIMSON RAMBLER one day said,
He did n't like the old homestead;
Thought he 'd travel, so he went
Over the wall on mischief bent.

GRANDDAD SAGE, the dear old man,
Says it is a splendid plan
For all young children to obey;
Says they did so in his day.

NOW LET the banners be unfurled,
To greet the fairest of the world;
Come Roses all, and pay your duty:
Madame the Queen, American Beauty!

GOLDEN-GLOW said "Well, I know,
I 'm just going to start and grow."
Liked it 'way up in the air —
Sent back word he 'd stay up there.

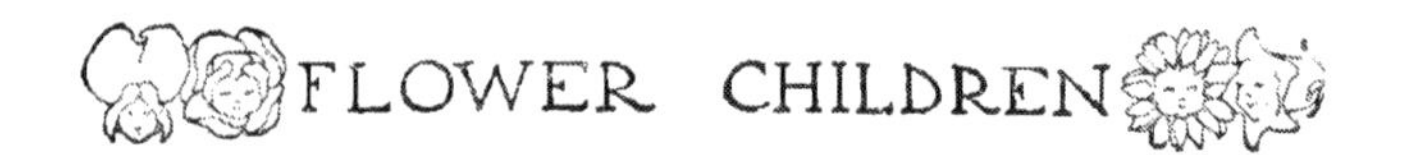

BACHELOR BUTTON, O, most shocking!
Found a hole in his silk stocking;
But he mended it so neatly,
Covered up the place completely.

JOLLY SUNFLOWER, big and yellow,
Said: "I 'm sure a lucky fellow.
To be small must seem so queer—
I get a splendid view from here."

STATELY Lady Hollyhock,
In a lovely colored frock,
Taught her children every day
Precisely what to do and say.

ZINNIA stands so very straight
Just inside the garden gate;
Sometimes single, sometimes double,
Never gives a bit of trouble.

BITTER-SWEET concluded she
Would live with some good, friendly tree;
Went to visit Madame Oak,
Stayed all winter, for a joke.

SAUCY LITTLE Black-eyed Susan,
When her mother caught her snoozin',
Rubbed her sleepy eyes and said
She guessed she 'd toddle off to bed.

NIGHTSHADE has a purple berry,
But he is very naughty, *very;*
Little children never should
Play with one who isn't good.

GENTIAN growing by the brook,
Bending low to get a look
At her pretty face so sweet,
Stepped too near and wet her feet.

SCARLET POPPY in the wheat,
Said she 'd like some grains to eat,
But when Head Wheat gave her some
She made believe 't was chewing-gum.

MULLEIN grew up rough-and-tumble.
He was Irish, very humble;
Still he was a jolly fellow,
With his funny head all yellow.

SIR THISTLE is a Scotchman bluff,
His manners are a trifle rough;
You find him everywhere you go;
He travels on the wind, you know.

WILD CUCUMBER said he guessed,
He 'd take a little trip out West,
Thought he 'd stay a year or two,
And maybe he 'd see something new.

BURDOCK and his family,
With the gardener don't agree;
But Burdock says if he 's your friend,
He 'll stick to you until the end.

CHINA ASTER thought he 'd do
The proper thing, and wear a queue;
But all his brothers laughed and said
He 'd better cut his hair instead.

CHRYSANTHEMUM is Japanese,
She 's a fine lady, if you please;
She comes to see us once a year,
About the time Thanksgiving 's here.

POISON IVY did n't know
Why every one disliked her so;
Made her feel so very sad
When people said she was so bad.

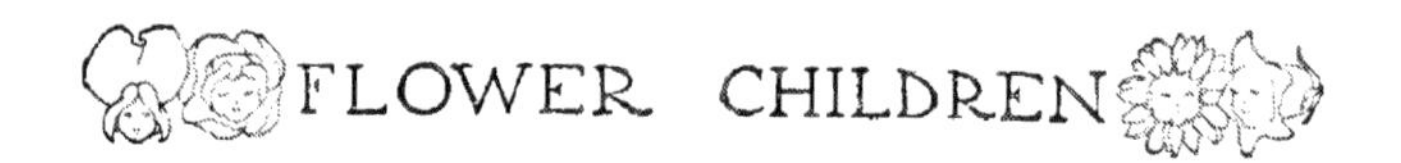

IN A SWEET velvet dress of red,
On Christmas Eve, Poinsettia said:
"I 'll hang my stocking up because
This is the night for Santa Claus."

EVER SEE a plant so jolly,
And good fellow-ish as Holly?
Makes no difference what's the weather,
He and Christmas come together.

Ingram Content Group UK Ltd.
Milton Keynes UK
UKHW021943270323
419267UK00005B/251